The Body Book to Navigating Puberty for Boys

A boy's guide to growing up and What they can expect

JAMEKA WATKINS

ISBN: 9798600150355

CONTENTS

FOREWORD

Puberty — it is a crazy time and occurs through a long process
But puberty is simply a series of natural changes that every child goes through.

This book contains guides for every guy to navigate puberty, positive guides to overcome puberty with confidence by respecting the body and all its changes.

It also helps parents who are forced to play a vital role in their sons' life to answer their everlasting questions "How to talk about puberty and body image?", "How to foster positive independence during puberty?", "How I can support my son during puberty?", … Wondering how to deal with your teenage son? Or how to raise teenage sons in general? Many parents are also seeking advice for understanding teenage boys.

Raising teenagers is not always easy and teenage boys' behaviors can be challenging. Rather, their actions and attitudes are the result of physiological and emotional turbulence during the adolescent years.

The book is an incredible tool to support parents for their sons when they begin the journey to puberty. The question of how to deal with your teenage son becomes easy to answer. What do kids really need? And how can we practice awareness to create authentic connection?

PUBERTY – ADOLESCENT MALE

How much will my teen grow?

The teenage years are also called adolescence. During this time, teens will see the greatest amount of growth in height and weight. Adolescence is a time for growth spurts and puberty changes. A teenager may grow several inches in several months followed by a period of very slow growth. Then they may have another growth spurt. Changes with puberty may happen slowly. Or several changes may occur at the same time.

It is important to remember that these changes will happen differently for each teen. Some teens may experience these signs of maturity sooner or later than others. And being smaller or bigger than other boys is normal. Each child goes through puberty at their own pace.

What changes will happen during puberty?

Sexual and other physical maturation that happens during puberty result from hormonal changes.

In boys, it is hard to know exactly when puberty is coming. There are changes that happen, but they occur slowly over a period of time rather than as a single event. There are certain stages of development that boys go through when developing secondary sex characteristics. Here is a brief overview of the changes that happen:

In boys, the first puberty change is the enlargement of the scrotum and testes. At this point, the penis does not enlarge. As the testes and scrotum continue to grow, the penis grows. The first growth of pubic hair produces long, soft hair that is only in a small area around the genitals. This hair then becomes darker and coarser as it continues to spread. The pubic hair eventually looks like adult hair, but in a smaller area. It may spread to the thighs and sometimes up the stomach.

The following changes may also happen to a boy as he goes through puberty:

Body size will increase. Sometimes the feet, arms, legs, and hands may grow faster than the rest of the body. This may cause a teen to feel clumsy. Some boys may get some swelling in the breast area. This is a result of the hormonal changes that are happening. This is common among teenage boys and is often a short-term or temporary condition. Talk with your son's healthcare provider if this is a concern. Voice changes may happen, as the voice gets deeper. Sometimes the voice may "crack" during this time. This is a temporary condition and will improve over time. Hair will start to grow in the genital area. Boys will also have hair growth on their face, under their arms, and on their legs.

As the puberty hormones increase, teens may have an increase in oily skin and sweating. This is a normal part of growing. It's important to wash daily, including the face. Acne may develop.

As the penis enlarges, the teen boy may begin to have erections. This is when the penis becomes hard and erect because it is filled with blood. This is due to hormonal changes and may happen when the boy fantasizes about sexual things. Or it may happen for no reason at all. This is normal.

During puberty, a boy's body also begins making sperm. Semen, which is made up of sperm and other body fluids, may be released during an erection. This is called ejaculation. Sometimes this may happen while the teen is sleeping. This is called a wet dream (nocturnal emission). This is a normal part of puberty. Once sperm is made and ejaculation happens, teen boys who have sex can get someone pregnant.

What does my teen understand?

The teen years bring many changes—not only physically, but also mentally and socially. During these years, teens increase their ability to think abstractly and eventually to make plans and set long-term goals. Each child may progress at different rates, and show a different view of the world. In general, the following are some of the abilities you may see in your teenager:
- Developing the ability to think abstractly
- Concerned with philosophy, politics, and social issues
- Thinking long-term
- Setting goals
- Comparing himself to his peers

Your teen's relationships with others

As your teenager begins to struggle for independence and control, many changes may happen. Here are some of the issues that your teen may experience during these years:
- He wants independence from parents.
- Peer influence and acceptance is very important.
- Peer relationships become very important.
- He may be in love.
- He may have long-term commitments in relationships.

WHAT BOYS WONDER ABOUT PUBERTY

For a guy, there is not just one event or sign that you are growing up. There are lots of them, including your body growing bigger, your voice changing, and hair sprouting everywhere. Most boys begin puberty between the ages of 9 and 14. But keep in mind that puberty starts when a boy's body is ready, and everyone grows at his own pace.

Why Are Girls Taller Than Me?

You might have noticed that some of the girls you know are taller than the boys. But you have probably noticed that out of the adults you know, most of the men are taller than the women. What's going on?

Well, girls get a head start on puberty — and growing taller — because they usually start these changes between the ages of 8 and 13. Most boys, on the other hand, don't begin until between the ages of 9 and 14. So that's why girls are often taller than boys during that time.

Most boys may catch up — and even grow taller than girls. But it is also important to remember that your genetics play a role in height. So if your mom and dad are tall, you are more likely to be tall. And if your mom and dad are kind of short, you may be short, too. But nothing is definite.

You have to wait and see how it turns out, but you can also talk to a doctor if you are concerned. Remember — not every adult male is tall. Many men who are considered "short" have gone on to have careers in the movies, the military, and even professional basketball!

There are not any exercises or magic pills to make you grow tall. But by being active and eating nutritious foods, you're helping your body grow up healthy, just the way it should.

When Will I Get Muscles?

During puberty, some boys might become worried about their bodies after seeing what some of their friends look like. For instance, lots of boys are concerned about their muscles. You may have already noticed some boys starting to get chest muscles (called the pectoralis muscles or pecs for short). Others may have broad shoulders (the deltoids). Other boys might still be slimmer and smaller.

Remember that puberty happens on its own schedule, so there is no rushing it if you're a little slower to develop muscles. Maybe you have considered lifting weights to help yourself get bigger. It's important to know that if you have not quite reached puberty, this will tone your muscles, but it will not build up any muscles yet.

Eating nutritious food and being active (like riding your bike, swimming, and playing sports) will help you be a kid who's strong and fit. In time, you will reach puberty and you can start building your muscles, too.

If you decide to try lifting weights, first let your doctor know you are interested. He or she may tell you to hold off on weightlifting for a bit or give you some advice on how to start. If your doctor discourages weightlifting, try some other ways to work out. Resistance bands, which are like big rubber bands, are a great way to help build your strength without putting too much strain on your muscles.

If your doctor recommends weightlifting, here are some tips:

Have a qualified coach or trainer supervise you. It is smart to have somebody show you the proper way to lift weights. This will help you gain strength and prevent injury.

Use lighter weights. Your coach or trainer can recommend the right amount. Lifting heavy weights can cause injuries and then you'll have to wait until you recover before you can work out again.

Do repetitions. It's better to lift a smaller amount of weight a bunch of times than to try to lift a heavy weight once or twice.

Rest. Let your body have a break at least every other day.

Do I Think About Girls Too Much or Not Enough?

There is this girl who lives in your neighborhood and you see her playing with her friends every afternoon when school is done. You get really hot and your palms sweat when she says "hi" to you. That night you go to bed and before you sleep, you have one last thought about her. Every day for the next few weeks you keep thinking about her. You might be wondering, "Why do I feel this way?" You just may have a crush.

Or perhaps your friend keeps talking about this one girl he thinks is so pretty. He goes on and on about how she tells funny jokes. He also tells you that he likes her. You think, "Why don't I feel or talk this way about a girl — am I supposed to?"

Every boy has his own likes and dislikes. And during puberty, some boys are very friendly with girls and others might be nervous about talking to girls. Thinking about someone you like is a normal process of puberty. And if you feel like you don't like any girls, that's fine, too. Eventually, you may find someone who makes you feel giddy inside. Only time will tell.

So why do you feel this way? The hormones in your body are becoming more active. As a result, you're starting to have more feelings. These feelings can confuse you and may leave you scared. This is natural because you are going through a new phase in your life.

Talking with a friend or an older person like your brother or sister might help you be less confused. Older people sometimes have more experience than you, so they can be good people to go to for advice.

What's Up with Body Hair?

Body hair really gets going during puberty. Some boys will start to notice hair growing on their face around the chin, on the cheeks, and above the lip. Also, hair grows on the chest, the armpits, and even down there in the pubic region. Remember that there's nothing to worry about because hair is just one of the body's many ways of telling you that you are on your way to manhood.

You're growing hair in new places because hormones are telling your body that it is ready to change. Some of the hormones that trigger this new hair growth come from your adrenal glands. Other hormones come from your pituitary gland (a pea-shaped gland located at the bottom of your brain). These pituitary hormones travel through your bloodstream and make your testicles ("balls") grow bigger and start to release another hormone called testosterone that also helps make your body start sprouting hair in your pubic area, under your arms, and on your face.

Boys don't really need to do anything about this new hair that's growing. Later, when you're a teen, and the hair gets thick enough on your face, you may want to talk with your parents about shaving.

Do I Smell?

You probably know what sweat is, but did you know that it's also called perspiration. How does it happen? Perspiration comes out of your skin through tiny holes called pores when your body gets hot.

Your body likes a temperature that is 98.6°F (37°C). If you get hotter than that, your body doesn't like it, so then your body sweats. The sweat comes out of the skin, then evaporates (this means it turns from a liquid to a vapor) into the air, which cools you down. Sometimes this sweat or wetness can be smelly and create body odor (sometimes called BO). During puberty, your hormones are working all the time, which explains why you sweat a lot and, well, sometimes smell.

What makes it smelly? The sweat is made almost completely of water, with tiny amounts of other chemicals like ammonia, urea, salts, and sugar. (Ammonia and urea are left over when your body breaks down protein.) Sweat by itself is not really smelly, but when it comes in contact with the bacteria on your skin (which everyone has) it becomes smelly.

But how can you keep yourself from being all sweaty and smelly? First, you can shower or bathe regularly, especially after playing sports or sweating a lot, like on a hot day. You can also use deodorant under your arms.
Deodorant comes in many good-smelling scents or you can use one that's unscented. Some deodorants come in a white stick that you can twist up. Lots of people put this on after showering or bathing before they put their clothes on. Otherwise, the white stick deodorants can leave white marks on your clothes. You can also choose a deodorant that's clear instead of white.
You can decide to wear a deodorant (which helps stops the smell) or a deodorant/antiperspirant (which helps stops the smell and the sweat). If you find these products are not working for you, talk with your doctor.

What About Erections?

An erection is what happens when your penis fills up with blood and hardens. The penis will become bigger and stand out from the body. Boys will start to notice erections occurring more often when they reach puberty. And they're perfectly normal.

An erection can happen at any time. You can get many in one day or none at all. It depends on your age, sexual maturity, level of activity, and even the amount of sleep you get.

An erection can happen even when you're sleeping. Sometimes you might wake up and your underwear or bed is wet. You may worry that this means you wet your bed like when you were little, but chances are you had a nocturnal emission, or "wet dream." A wet dream is when semen (the fluid containing sperm) is discharged from the penis while a boy is asleep. Semen is released through the urethra — the same tube that urine (pee) comes out of. This is called ejaculation.

Wet dreams occur when a boy's body starts making more testosterone. This change for boys is little bit like when a girl gets her period. It's a sign a boy is growing up and the body is preparing for the day in the future when a man might decide to be a father. Semen contains sperm, which can fertilize a woman's egg and begin the process that ends with a baby being born.

Although some boys might feel embarrassed or even guilty about having wet dreams, a boy can't help it. Almost all boys normally experience them at some time during puberty and even as adults.

But if you ever have pain or a problem with your penis or testicles, it is important that someone take you to the doctor. You may think "Man, I do not want to go to the doctor for that!" But it is best to get problems like this checked out — and your doctor will not be embarrassed at all. It is a doctor's job to help you take care of your body — even that part.

THE 5 STAGES OF PUBERTY IN BOYS

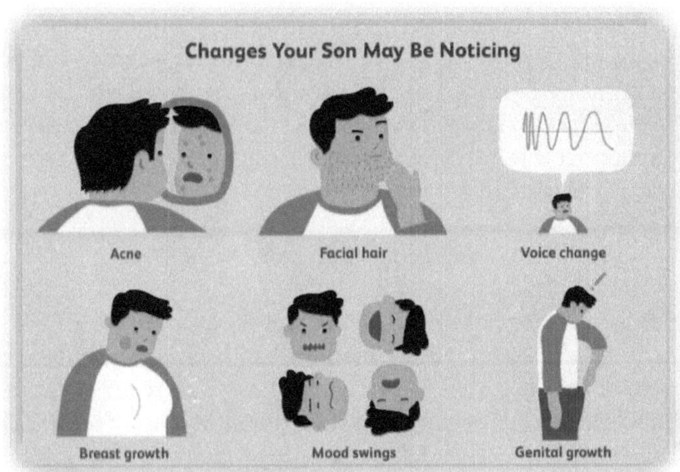

Before you know it, your little boy has become a young man.
There are five stages of puberty that boys go through, but keep in mind that the age at which each boy goes through them can vary widely.

Signs of Puberty in Boys

Boys mature a little slower than girls. For boys, puberty begins at age 11 on average, although starting as early as age 9 or as late as age 14 is still considered normal. Some boys mature faster than their peers, and some physical changes may be more gradual than others.

A number of these physical changes are very personal. As a parent, you may not notice them, but your son will. Some of these may be embarrassing experiences for him and he will likely keep much of this private.

Body Shape

Externally, you may notice your son's body begin to grow, but just before that happens, he may put on a little weight and look like he is all arms and legs. Next comes a growth spurt in height, often around the age of 13. His shoulders will broaden and his muscles will develop more definition too. He will become noticeably stronger and can take advantage of that by beginning a regular workout routine.

Sweating, Hair, and Acne

Personal hygiene is probably one of the biggest changes for young boys. It may have been hard to get him to wash his hands or take a shower, but now he will need to pay attention to these things as he starts to sweat more and develop body odor. He may soon come to you and ask about shaving the peach fuzz from his face or ask about antiperspirants. His hormones will produce more oil on his skin and he may be prone to acne break-outs. This is the perfect time to introduce him to good skin care routines.

Penis and Testicle Growth

The first sign of puberty actually begins with the growth of your son's testicles and scrotum, which will more than double in volume.[6] His penis and testicles will begin to grow as he enters puberty too, as will his pubic hair. The penis begins by growing in length and width. You can find more details on this growth in the Tanner stages section below.

Around one-third of boys have tiny pearly bumps, called papules, on their penises. These bumps look like pimples and are normal and harmless, though they are permanent.

Nocturnal Emissions and Erections

As your son develops, he may begin to have nocturnal emissions, or "wet dreams," in which he ejaculates at night. This can occur with or without a sexual dream and is completely normal. Talking to your son about nocturnal emissions before they happen is helpful so he knows what to expect and that he hasn't accidentally wet the bed. Let him know that it's just another part of puberty and that it'll go away in time.

Involuntary erections are another big part of male puberty and they can occur at any time, for absolutely no reason at all. Explain to your son that this may happen for a while, and he will likely have little control over it, but it will get better as he gets older.

Voice Change

Your son's voice will change around the time that his growth spurt has begun to slow down a bit. This occurs because his vocal chords and voice box (larynx) gain mass too. Before his voice changes completely, it may crack and soar, going from high to low quickly.[9] This can be embarrassing for him, so be mindful of this.

Breast Growth

When your son is first in puberty, his breast tissue may swell a bit for a year or two as some of his hormones change into estrogen. For the majority of boys, this is temporary and not excessive, though, in some boys, it can be more obvious, especially if they are overweight. If your son's breast area seems excessively swollen or the swelling happens before puberty or later in puberty, see your healthcare provider. There could be a medical problem that's causing this swelling rather than hormones.

Mood Swings

Like girls, boys can also have mood swings thanks to the hormonal, physical, and emotional changes they are experiencing. Be patient and understanding; this, too, shall pass.

Tanner Stages of Sexual Development

Teen boys will develop physically in certain stages, often called Tanner stages. Your pediatrician or family health care provider can determine what stage your teen is at and if it's expected for his age. The Tanner stages, along with approximate age ranges, include:

Sexual Maturity Rating 1: (The pre-puberty stage) The testes are small and the phallus (penis) is child-like. There is no pubic hair.

Sexual Maturity Rating 2: (From 10 years old to 15 years old) The testicles grow in volume and size. The penis has no to slight enlargement. The scrotum becomes reddened, thinner, and larger. A few pubic hairs become visible and they are long, straight, and slightly dark.

Sexual Maturity Rating 3: (From 10 years old to 16 years old) The testes continue to grow in volume and size. The penis becomes longer. The scrotum continues to enlarge. Pubic hairs become darker and curlier and more of them appear.

Sexual Maturity Rating 4: (From 12 years old to 17 years old) The testicles continue to grow. The penis continues to grow in length and now becomes thicker. The scrotum grows larger and also darkens. Pubic hair is coarse, thicker, and curly like adult hair, though there are fewer hairs than an adult has.

Sexual Maturity Rating 5: The testicles are of an adult size (greater than 20 ml in volume). The scrotum and penis are of adult size and form. The pubic hair is of normal adult distribution and volume.

Talking to Your Son

Your little boy is growing up and this also means that he may open up to you less often. It's common for teenage boys to become less talkative and withdraw from their parents. Keep the lines of communication open and talk to your son about the changes he's experiencing. Stay connected to his interests and talk to him about sports, school, or whatever he enjoys. This will help him feel comfortable about coming to you when he needs to talk about something important.

Delayed Puberty

If your son has not started puberty by the age of 14, which means that his testicles and penis have not started to grow yet, this is considered delayed puberty. The most common cause is called constitutional delayed puberty. Most boys who are constitutionally delayed are totally healthy and will go through puberty eventually. More than two-thirds of boys inherit this from one or both of their parents who also started puberty late. In boys, this can be defined as having no increase in testicle size by the age of 14 years old or continuing undergo puberty for more than five years after the start. In girls, delayed puberty is starting menstruation after the age of 16 years. The majority of boys who are constitutionally delayed are also short compared to other boys their age, but this is just because they have not had their growth spurt yet.

If your son has a chronic illness like sickle cell disease, inflammatory bowel disease, or cystic fibrosis, puberty may also begin later than normal.

A small number of boys have a condition called isolated gonadotropin deficiency (IGD), which means that they do not produce adequate amounts of the hormones LH and FSH. This condition typically begins at birth and is typically treated with testosterone injections.

An even smaller number of boys have something going on with their testicles that is causing puberty to be delayed. Testosterone is the main treatment for issues of this sort.

A Word from Very well

If you have questions or concerns about how your son is progressing through puberty, talk to his healthcare provider. Your doctor can determine if your teen is growing and developing as expected and help you understand the biology that's at work. In the case of suspected delayed puberty, it is possible that your son's penis and testicles have gradually started to enlarge and he just hasn't noticed. Your doctor can tell with a physical exam and can run some tests on your son's hormones to see if there are any problems.

PHYSICAL DEVELOPMENT IN BOYS: WHAT TO EXPECT

Puberty is a crazy time and occurs through a long process, beginning with a surge in hormone production, which in turn causes a number of physical changes.
Every person's individual timetable for puberty is different.
This chapter is an overview of some physical changes boys can expect during these years.

Enlargement of the Testicles and Scrotum

A near doubling in the size of the testicles and the scrotal sac announces the advent of puberty. As the testicles continue to grow, the skin of the scrotum darkens, enlarges, thins, hangs down from the body and becomes dotted with tiny bumps. These are hair follicles. In most boys, one testicle (usually the left) hangs lower than the other.

Pubic Hair

Fueled by testosterone, the next changes of puberty come in quick succession. A few light-colored downy hairs materialize at the base of the penis. As with girls, the pubic hair soon turns darker, curlier and coarser in texture, but the pattern is more diamond-shaped than triangular. Over the next few years it covers the pubic region, then spreads toward the thighs. A thin line of hair also travels up to the navel. Roughly two years after the appearance of pubic hair, sparse hair begins to sprout on a boy's face, legs, arms and underarms, and later the chest.

Changing Body Shape

A girl's physical strength virtually equals a boy's until middle adolescence, when the difference between them widens appreciably. Boys tend to look a little chubby and gangly (long arms and legs compared to the trunk) just prior to and at the onset of puberty. They start to experience a growth spurt as they progress further into puberty, with the peak occurring during the later stages of sexual maturation. Body proportions change during this spurt, as there is rapid growth of the trunk, at the legs to some extent too. Boys continue to fill out with muscle mass long after girls do, so that by the late teens a boy's body composition is only 12 percent fat, less than half that of the average girl's.

Penis Growth

A boy may have adult-size genitals as early as age thirteen or as late as eighteen. First the penis grows in length, then in width. Teenage males seem to spend an inordinate amount of time inspecting their penis and covertly (or overtly) comparing themselves to other boys. Their number-one concern? No contest: size.

Most boys do not realize that sexual function is not dependent on penis size or that the dimensions of the flaccid penis don't necessarily indicate how large it is when erect. Parents can spare their sons needless distress by anticipating these concerns rather than waiting for them to say anything, since that question is always there regardless of whether it is articulated. In the course of a conversation, you might muse aloud, *"You know, many boys your age worry that their penis is too small. That almost never turns out to be the case."* Consider asking your son's pediatrician to reinforce this point at his next checkup. A doctor's reassurance that a teenager is *"all right"* sometimes carries more weight than a parent's.

Boys' preoccupation with their penis probably will not end there. They may notice that some of the other guys in gym have a foreskin and they do not, or vice-versa, and might come to you with questions about why they were or weren't circumcised. You can explain that the procedure is performed due to parents' choice or religious custom.

"What Are These Bumps On My Penis?"

About one in three adolescent boys have penile pink pearly papules on their penis: pimple-like lesions around the crown, or corona. Although the tiny bumps are harmless, a teenager may fear he's picked up a form of sexually transmitted disease. The appropriate course of action is none at all. Though usually permanent, the papules are barely noticeable.

Fertility

Boys are considered capable of procreation upon their first ejaculation, which occurs about one year after the testicles begin to enlarge. The testicles now produce sperm in addition to testosterone, while the prostate, the two seminal vesicles and another pair of glands (called Cowper's glands) secrete fluids that combine with the sperm to form semen. Each ejaculation, amounting to about one teaspoonful of semen, contains 200 million to 500 million sperm.

Wet Dreams & Involuntary Erections

Most boys have stroked or rubbed their penises for pleasure long before they are able to achieve orgasm—in some instances, as far back as infancy. A child may consciously masturbate himself to his first ejaculation. Or this pivotal event of sexual maturation may occur at night while he's asleep. He wakes up with damp pajamas and sheets, wondering if he'd wet the bed.

A nocturnal emission, or "wet dream," is not necessarily the culmination of a sexually oriented dream.

What parents can do to help:

- Explain to your son that this phenomenon happens to all boys during puberty and that it will stop as he gets older.

- Emphasize that a nocturnal emission is nothing to be ashamed of or embarrassed by.

- Note that masturbation is normal and harmless, for girls as well as boys, as long as it is done privately.

Erections, too, are unpredictable during puberty. They may pop up for no apparent reason—and seemingly at the most inconvenient times, like when giving a report in front of the class. Tell your teen there is not much he can do to suppress spontaneous erections (the time-honored technique of concentrating on the most unsexy thought imaginable does not really work), and that with the passage of time they will become less frequent.

Voice Change

Just after the peak of the growth spurt, a boy's voice box (larynx) enlarges, as do the vocal cords. For a brief period of time, your son's voice may "crack" occasionally as it deepens. Once the larynx reaches adult size, the cracking will stop. Girls' voices lower in pitch too, but the change is not nearly as striking.

Breast Development

Early in puberty, most boys experience soreness or tenderness around their nipples. Three in four, if not more, will actually have some breast growth, the result of a biochemical reaction that converts some of their testosterone to the female sex hormone, estrogen. Most of the time the breast enlargement amounts to a firm breast bud of up to 2 inches in diameter under the nipples. Occasionally, this may be more extensive, resulting in profound "gynecomastia." Overweight boys may have the appearance of pseudo-gynecomastia due to excess fatty tissue on the chest wall.

As you might imagine, this development can be troubling for a child who is in the process of trying to establish his masculinity. If your son suddenly seems self-conscious about changing for gym or refuses to be seen without a shirt, you can reasonably assume that he's noticed some swelling in one or both breasts. (One particularly telltale sign: wearing a shirt to go swimming.) Boys are greatly relieved to learn that gynecomastia usually resolves in one to two years. *"Thanks for telling me! I thought I was turning into a girl!"* is a common reaction. There are rare instances where the excess tissue does not subside after several years or the breasts become unacceptably large. Elective plastic surgery may be performed, strictly for the young patient's psychological well-being.

Gynecomastia warrants an evaluation by a pediatrician, especially if it arises prior to puberty or late in adolescence, when the cause is more likely to be organic.

A number of medical conditions can cause excessive breast growth, including:

- Endocrine tumors
- An adrenal disorder (congenital adrenal hyperplasia)
- A chromosomal disorder
- Liver disease
- Rare genetic disorders

Breast development may also be a side effect of various drugs, including certain antidepressants, anti-anxiety medications, anti-reflux medications, or due to exposure to external sources of estrogen or estrogen precursors, including ingested soy, and plant estrogen in lotions and/or personal care products, such as lavender or tea tree oil applied to the skin. There may be other possible environmental sources, some of which are under investigation, such as certain plastic containers.

PUBERTY BOYS QUIZ

Have you hit Puberty, boys?

1. I have pubic hair.

- True
- False

2. I have armpit hair.

- True
- False

3. I have sexual thoughts about members of the opposite gender (unless you are gay).

- True
- False

4. I have been growing taller.

- True
- False

5. I sometimes feel withdrawn from my parents.

- True
- False

6. My penis size has increased.

- True
- False

7. I have been feeling unfamiliar emotions lately.

- True
- False

8. I am sweating a lot.

- True

False

9. My chest has been swelling.

○ True

○ False

10. I care about my appearance.

○ True

○ False

11. Have you become a bit sweatier/smelly than normal?

○ No. I'm as fresh as a daisy!

○ Kind of. wear deodorant

○ I haven't used so much deodorant!

○ Just a little

○ Yes. Need extra deodorant

12. Have you gotten a lot taller?

○ 1/2 inch

○ 3-4 inches or more

○ Nope. I'm still short

○ 2 inches to 2 1/2 inches

○ 1 inch to 1 1/2 inches

13. Has your voice changed? (Includes cracking, getting deeper)

○ Nope! I'm still a kid!

○ Yes! (Voice crack)

○ Yeah, there's a little bit going on

○ Some cracking

14. Has your penis/testicles gotten longer? (I know, its awkward)

A little.

Kind of

OH MY GOD, ITS SO LONG!

Yes

Nope!

15. How old are you? (Most boys start puberty between 12-18.)

15-16

I am 12-13.

14-15

I am less than 12.

Over 16. I probably have hit puberty.

16. Have you developed abs/muscles?

I'm STRONG! *Kisses muscles*

Kind of, not really

Nope!

I'm a bit stronger than usual.

CAN THEY COME BIGGER? THEY'RE HUUUUGE!

17. What's your favorite color? (Doesn't count)

Red

Other

Green

Blue

Orange

18. What's your favorite emoji (Also doesn't count)

○ POOP

○ Other

○ I hate emoji!

○ Heart eyed emoji

○ Cool emoji

19. Do you feel more attracted to girls/guys?

○ Yes.

○ Kind of

○ A little

○ I. NEED A PARTNER.

○ No

20. Do you have any hair on your penis?

○ Little bit

○ Nope!

○ Kind of

○ Yes

○ My penis is covered in curly black hair

HOW TO DEAL WITH YOUR TEENAGE SON

What do kids really need?
And how can we practice awareness to create authentic connection?

The Basics of How to Deal with Your Teenage Son

A few keys for how to deal with your teenage son: Communicate with him often, do things together as much as possible, and give him unconditional love.

Of course, all that is easier said than done. However, understanding teenage sons may be less of a problem when you're familiar with the process of adolescent development.

Remember, teen boys are growing in all sorts of ways. Therefore, you can offer compassion and support. Moreover, trying a few new approaches to parenting teen boys can help.

First, let's look at the growth process that's taking place in a teenage boy's body.

What's Happening in a Teenage Son's Body and Brain?

Teenage boy behavior is controlled in large part by the many hormonal and biological changes that occur during puberty. Puberty in boys starts between 10 and 14 years old. And teenage boys are physically mature around age 15 or 16. Hence, boys grow taller, develop larger muscles, and get deeper voices.

Along with physical changes, teen boys experience emotional and behavioral changes. Teen puberty is an exciting time, full of new emotions and feelings. Therefore, it affects teenage boy behavior as well as their interest in sex and relationships.

Furthermore, the adolescent brain is still developing throughout the teenage years. Moreover, the area of the brain that's responsible for judgment and decision-making remains under construction. This area, the prefrontal cortex, doesn't fully mature until the mid-20s. Hence, teen boys are more susceptible to shifting impulses and emotions during teen puberty. How to deal with your teenage son gets complicated.

Teen Boys and Risky Behavior

When parents talk about a "problem teenage son," they are often referring to their teen's risky behavior. Teenage boys may engage in a number of risky behaviors, including:

- Reckless driving—such as speeding, or texting while driving
- Unprotected sex
- Substance abuse
- Binge drinking
- Cigarette smoking
- Behavior that may lead to violence or injury—e.g., fighting, carrying weapons, or participating in unsafe recreational activities.

There are many reasons why teenage boys are drawn to **risk-taking behaviors**. For one, the part of the brain that controls impulses are not mature yet. Moreover, taking risks can be a misguided way for teens to strike out on their own and feel independent. In addition, peer pressure can be a factor. Plus, external stresses can push teenage boys toward risky behaviors to let off steam.

Teenage Behavior Management Strategies

When figuring out how to deal with your teenage son's risky behavior, parents need to create clear limits and effective consequences. Hence, when navigating how to deal with your teenage son, and putting teenage behavior management strategies in place, take a direct approach.

Five Keys for Dealing with Teenage Boy Behavior:

Set limits. First, parents and teen boys agree to **set boundaries** and rules that both agree on. The rules are based on shared values about staying safe and keeping harmony in the family.

Write it down. Furthermore, families might consider drafting a written agreement. Therefore, the guidelines and boundaries are clear to everyone.

Agree on consequences. Next, parents and sons agree on age-appropriate consequences that will go into effect if the rules are broken. For example, a consequence might be loss of car privileges or an earlier curfew. Moreover, the consequence should be age-appropriate.

Invoke restitution. In addition, parents and teen boys can use a consequence known as restitution or restoration. Hence, teens help make a situation better after violating the shared contract. For example, if they get a speeding ticket, they pay it on their own. Or they take steps to repair a relationship with a sibling after a fight. As a result, a teen can earn back parents' trust.

Avoid severe punishment. However, severe punishment is not the best approach for dealing with your teenage son. In fact, punishment can make things worse. Teenage boys may feel rejected and resentful. Hence, they may withdraw further from their parents.

Research shows that teenage sons do better when their parents remain warm, open, and supportive, while also setting firm boundaries.

Self-Care in Teenage Boys

Teenage boys are notorious for poor self-care. That is, many teen boys do not sleep enough. In addition, they eat junk food and drink beverages high in sugar. Furthermore, they may not shower or wash on a daily basis. And they might neglect physical exercise—sometimes in favor of screen time.

In part, teen boys' poor self-care comes from being self-conscious about their changing bodies. The physical development that comes with puberty can trigger body-image and self-esteem issues. Hence, teens don't know how to deal with body odor, acne, and/or oily hair—all of which can come with puberty.

This teenage boy behavior can be helped by setting routines around healthy eating, exercise, and good sleep hygiene when their son is young. In addition, younger adolescent boys might need basic information about grooming and self-care during puberty. If parents aren't comfortable sharing this information themselves, they might instead find a book or pamphlet. Subsequently, they can leave it in their son's room for him to read when he's ready. No need to make a big deal about it—remember, teen boys are easily embarrassed. Later, at an appropriate time, ask if he has any questions about what he read.

The Effect of Video Gaming on Teenage Boys

According to a national survey from the Pew Internet & American Life Project, 99 percent of boys aged 12 to 17 play video games. Moreover, they play often. In fact, half of the teens surveyed said that they had played a video game the previous day.

In addition, a recent study found that screen time can increase symptoms of teenage ADHD (Attention Deficit Hyperactivity Disorder). Hence, teenage boys who play video games and use social media for hours each day are at risk. For some parents, this might provide additional challenges around how to deal with your teenage son.

The study revealed that teenagers who used digital media often were more than twice as likely to develop symptoms of ADHD. Hence, teenage boys who play video games for hours may experience increased impulsivity, a lack of focus, and an inability to concentrate for a sustained period of time. And most parents find these behaviors difficult to handle or think they have a problem teenage son.

Setting Limits on Screen Time for Teen Boys

What can parents do to help their sons unplug? When kids are younger, parents can set time limits. But that becomes harder to enforce as teens get older.

Therefore, parents need to carve out times with no screens allowed, such as meals and family activities. And they need to model this behavior by staying off their own phones and other devices.

Moreover, after a certain hour of the evening, parents can turn off the Wi-Fi so teens can't use the Internet. Teens should know this isn't a punishment. Rather, screen time disturbs sleep. So it's important to unplug an hour or more before bed to allow the nervous system's relaxation response to kick in.

In addition, just as with self-care, good habits stick best when they are instilled early. Parents can help teenage boys develop habits that take them away from screens:
- Spending time outdoors
- Connecting with friends IRL (in real life)
- Playing sports, running, or other physical activities
- Volunteering
- Creative expression, such as playing an instrument or drawing.

Relationships Between Mothers and Sons

As boys grow into teens, their relationships with their mothers can become a little bumpy. That's because teenage sons and mothers need to create appropriate boundaries. For teen boys, part of maturing is becoming more independent from their mothers. Hence, a teenage son being disrespectful to his mother is a sign that he is pulling away to learn how to care for himself. Fathers often connect with their teenage sons by doing things together. However, mothers and teenage sons sometimes have fewer interests in common. Therefore, mothers need to find ways to spend time with their teenage sons while also giving them their space, this is an important part of understand teenage sons and their needs.

A study supported by the National Institutes of Health looked at the impact of the mother-son relationship on teen behavior. As a result, researchers found that boys who experienced a lot of conflict with their mothers were more likely to engage in delinquent behavior as teens. But boys who had a close relationship with their mothers were more likely to have a better relationship with their best friends during the teen years.

Therefore, the study concluded that successfully adapting to the transitions of childhood and adolescence requires high levels of closeness and openness between parents and children. Moreover, minimizing conflict is also essential. And good teen-parent relationships set kids up to create their own successful relationships outside the family.

Communicating with Your Teenage Son

Teen boys are not known for their skill in communication. Often, teenage sons find it difficult to put their emotions into words. Understanding teenage sons begins with knowing they may not feel comfortable sharing their innermost thoughts with their parents.

As a result, parents can get frustrated and feel ignored. Instead, try the following approaches.

Keep it short and sweet. If you have something you need him to know, offer a series of clear points. Subsequently, let him respond to each.

Don't overdo the eye contact. While eye contact is often recommended for effective communication, that doesn't hold true for dealing with your teenage son. Instead, it might overwhelm or intimidate him. For that reason, driving in the car together can be a good time for talking.

Talk while you are in action. Many teen boys find it easier to communicate when they're doing something else at the same time. So have your chat while playing a game, taking a hike, or preparing dinner together.

Stay calm. When assessing how to deal with your teenage son, don't let your emotions get the upper hand. Showing anger or frustration may drive him deeper into his shell. As a result, he will be less likely to come to you for support.

Give him time to process. Many teenage boys need a few hours or even days to think about important conversations. Therefore, don't be disappointed if your teenage son doesn't change his behavior or attitude right away. Let him take in the information and then process it in his own time.

Never Underestimate the Power of Parents

Sometimes parents might feel that their teenage son has no interest in them. But parents should not let that fool them. How to deal with your teenage son is stay involved, no matter what.

In summary, evidence clearly points to the continued importance of the attachment between parent and teenage sons. As a result, this ongoing relationship supports teen mental health and decreases substance abuse. Moreover, healthy teen-parent relationships help adolescents grow into strong, independent young adults.

PARENTING YOUR CHILD THROUGH PUBERTY

Puberty is a time of great change for your child — and for you as a parent too.

Puberty brings lots of changes for your child – and for you as a parent too. Your child is transitioning from child to adult, and you may feel uncertain about how best to support them through the physical, psychological and emotional changes this brings.

Never fear, there's plenty you can do to help your child. One of the best ways is to just be reassuring.

Puberty is simply a series of natural changes that every child goes through. Some kids struggle with the changes, while others sail through puberty without concern. Only a small percentage of children experience extreme turmoil during this phase of their development.

Puberty can also be exciting and special, and as your child's parent, you are in the ideal position to help them through it.

What to expect during puberty?

You can read more detailed articles on puberty and the teenage years in general, but this short summary gives you an idea of what to expect.

The changes of puberty are physical, sexual, social and emotional. Puberty starts when changes in your child's brain cause sex hormones to be released in the:

- ovaries (usually around age 10 or 11, but can range from 8 to 13 years), or
- testes (usually around age 11 to 13, but can range from 9 to 14 years).

You cannot predict how long your child will go through puberty. It may be anywhere from 18 months up to 5 years. Genetic, nutritional and social factors determine when puberty starts and for how long it runs.

During puberty, most children will experience:

- oily skin (**acne** is possible)
- oily hair, possibly requiring frequent washing
- increased perspiration and body odor (frequent showering and deodorant help)
- a growth spurt (of around 11 cm a year in girls and up to 13 cm a year in boys). Teens continue to grow about 1–2 cm a year after this main growth spurt. Some body parts (such as head and hands) may grow faster than limbs and torso. The body eventually evens out.

Boys will experience:

- growth of the penis and testes (testicles). Sometimes the growth of the testes is uneven (that is, one testis grows faster than the other). This is not something to worry about
- growth of pubic, underarm and facial hair
- the start of testosterone production, which stimulates the testes to produce sperm
- the start of erections and ejaculation
- growth of the larynx or voice box – the voice 'breaks' and eventually deepens. Voice variations are normal and will settle in time.

What to expect socially and emotionally:

Mood changes and energy level variations are normal parts of puberty, as are swings between feeling independent and wanting parental support.

Your child will want to establish their own identity, which may include new friendships and experiences. If this happens, they will encounter challenges about how to manage current friendships. They may also start to explore their sexuality and may go on dates and start developing romantic relationships.

Puberty and adolescence is a time for children to become more independent (such as getting themselves to and from school). They may also be looking for more responsibility, such as taking on a leadership position at school, or finding a part-time job.

Your child may also be sensitive about how they look and their new body changes. Privacy and personal space may become very important to them. They may alternate between feeling self-conscious about themselves one day, to feeling 'bullet proof' the next.

These social and emotional changes show your child is forming their own identity and learning how to be an independent adult. They are developing their decision-making skills and learning to recognize and understand the consequences of their actions.

Teenagers and social media

Social media use is common among teenagers. It has a range of benefits (such as connecting with friends, feeling less isolated, exposure to new ideas) and risks (such as cyberbullying, sexting and spending too much time online) associated with its use.

How you can support your child during puberty

One of the best strategies during your child's puberty is reassurance. Explain that puberty is an exciting time that means adulthood is approaching.

Try to show compassion for the changes they are experiencing and reassure them the changes are normal – and many will pass. Of course, if you're concerned about your child's development, talk to your healthcare professional.

Puberty is also a time when role-modelling body acceptance is really valuable. Your child will compare their body to those of their friends, and may feel worried about their own development. The best thing you can do is show understanding and explain bodies come in all shapes and sizes. Modelling a healthy lifestyle will also help your child.

Be accepting of your child's need for privacy, and that your child may be exploring their body through masturbation. Always knock before entering their room.

If your child is early or late to puberty, be understanding and offer lots of reassurance and support. They may feel embarrassed but let them know everybody develops at their own pace.

You may also find it useful to keep the following tips in mind:

- Praise your teenager for their efforts, achievements and positive behavior.
- Put yourself in your child's shoes, and try to see their behavior for what it often is: your child struggling to become an individual.
- Try to stay calm during angry outbursts from your child. Wait for your child to cool down before talking about the problem.
- Stay interested and involved, and be available if your child wants to talk.
- Chat to your partner or other parents of teenagers. Sharing concerns and experiences can ease the load.
- Try to support your child in their self-expression, even if some of it seems odd to you, such as an extreme haircut or offbeat clothing choices.

- Try to tolerate long periods of time spent on personal care, such as hours in the bathroom, but chat to your child about reasonable family time limits.
- Talk to your child about any permanent changes they want to make to their body, such a tattoos and piercings, and discuss temporary alternatives, such as henna (removable) tattoos.
- If your child has acne, talk to them about how they feel about it. If it is bothering them, ask if they would like to see a doctor. Your doctor may refer your teenager to a skin specialist or dermatologist.

How you can support your son during puberty

Helping your son through puberty is mostly about reassurance. Reassure your son that testes develop unevenly, and it is common for one to be lower than the other. If your son's testes are very small or not both in the scrotum, see your GP.

You may also need to reassure your son that penis size does not affect sexual functioning, and that erect penises are usually very similar in size. Every boy develops in his own time. Ejaculating during sleep (sometimes called a wet dream) and spontaneous erections are both normal.

If your son experiences breast growth or tenderness, he may be concerned. Again, reassurance is the key. Any tenderness is likely to settle once his chest widens. If your son feels small or too thin for his age, reassure him he will grow in time.

Remember, you know your child best. If anything about their development concerns you, see your GP.

How to talk about puberty and body image

The best time to talk about puberty with your child is before it begins. Take an open and relaxed approach to chatting with your child.

Use the correct terms for body parts so your child learns the right words and is comfortable using them when talking about their body. They need to know their body parts are normal and natural, with words to match.

You may like to open a conversation by asking whether your child has learned about puberty at school and what they have been taught.

Convey facts in the conversation, such as 'Every kid goes through these changes, but not always at the same time. Have you noticed that?' And talk about your values too: for instance, you may choose to say that you think a behavior such as masturbation is a normal way to handle sexual feelings.

Pick a time to talk when there are no distractions, and don't be worried if your child doesn't want to share everything with you. They may prefer to talk to your family doctor or a counsellor.

How to foster positive independence during puberty

It is normal for your child to want more independence – but still need your support – during puberty or teen years. They may take risks as they explore their boundaries.

As a parent, you may be worried about your child's safety, and find yourself arguing with them about their push for independence. Try to stay calm and work through the issues with your child. Communicate openly, and make sure your child knows you are there for them. Stay available, because being accessible is the best way to find out what your child is doing and to help keep them safe.

Talk to your child about making good decisions, and your family's values. Ask your child to tell you where they are and what they're doing.

How to look after yourself at this time

It's important that you look after yourself during this potentially challenging time in your child's development. Trust in your skills as a parent – and talk to others or read up on the subject so you feel confident in guiding your child through it.

Puberty is the beginning of your child's transformation into an adult. Take some time to accept that your child, and your role as parent and your family dynamic, is changing.

You may also need to accept that you will not have total control over your child's choices and life direction once they are a young adult. It may help to trust that you have done your best as a parent and trust in your young person. But if your child makes new friendships that lead to activities that concern you, such as violence or drug taking, you may feel particularly stressed. In these times it may be useful to seek the advice of a family counsellor or a service which offers parenting advice as well as relationship education programs.

Stay available and caring. Let your child know you are there for them, no matter how old they are. Take some time for yourself to reduce stress, and look after your own needs if this time is particularly challenging.

Some tips for ways to take care of yourself:

- Prepare a weekly family plan, so you know what people are doing and where they need to be. Include some fun family rituals, like Saturday night cards, or maybe a weekly walk or bike ride. Do not forget to schedule some time for yourself.
- Nurture your relationship with your partner. Remember, they're facing many of the same challenges that you are. A regular date night in your family schedule can work wonders.
- Use your support networks, like grandparents, other family members and friends. What child (including your teenager) doesn't enjoy being spoiled by a doting grandparent? You could also share carpooling or supervision duties with friends.
- Ask the kids to help out with household chores. Your child will learn some new skills, gain some new responsibility, and it will lighten the load for you as parents.
- Stay positive and keep things in perspective.

COMMUNICATING WITH YOUR PUBERTY BOYS

Keeping the parent-child relationship strong during a tricky age

The teenage years have a lot in common with the terrible twos. During both stages our kids are doing exciting new things, but they're also pushing boundaries (and buttons) and throwing tantrums. The major developmental task facing both age groups is also the same: kids must pull away from parents and begin to assert their own independence. No wonder they sometimes act as if they think they're the center of the universe.

This makes for complicated parenting, especially because teens are beginning to make decisions about things that that have real consequence, like school and friends and driving, not to speak of substance use and sex. But they aren't good at regulating their emotions yet, so teens are prone to taking risks and making impulsive decisions.

This means that having a healthy and trusting parent-child relationship during the teenage years is more important than ever. Staying close isn't easy, though. Teens often aren't very gracious when they are rejecting what they perceive to be parental interference. While they're an open book to their friends, who they talk to constantly via text messages and social media, they might become mute when asked by mom how their day went. A request that seemed reasonable to dad may be received as a grievous outrage.

If this sounds familiar, take a deep breath and remind yourself that your child is going through his terrible teens. It is a phase that will pass, and your job as parent is still vitally important, only the role may have changed slightly. Here are some tips for navigating the new terrain:

1. Listen. If you are curious about what's going on in your teen's life, asking direct questions might not be as effective as simply sitting back and listening. Kids are more likely to be open with their parents if they don't feel pressured to share information. Remember even an offhand comment about something that happened during the day is her way of reaching out, and you're likely to hear more if you stay open and interested — but not prying.

2. Validate their feelings. It is often our tendency to try to solve problems for our kids, or downplay their disappointments. But saying something like "She wasn't right for you anyway" after a romantic disappointment can feel dismissive. Instead, show kids that you understand and empathize by reflecting the comment back: "Wow, that does sound difficult."

3. Show trust. Teens want to be taken seriously, especially by their parents. Look for ways to show that you trust your teen. Asking him for a favor shows that you rely on him. Volunteering a privilege shows that you think he can handle it. Letting your kid know you have faith in him will boost his confidence and make him more likely to rise to the occasion.

4. Don't be a dictator. You still get to set the rules, but be ready to explain them. While pushing the boundaries is natural for teenagers, hearing your thoughtful explanation about why parties on school nights aren't allowed will make the rule seem more reasonable.

5. Give praise. Parents tend to praise children more when they are younger, but adolescents need the self-esteem boost just as much. Teenagers might act like they're too cool to care about what their parents think, but the truth is they still want your approval. Also looking for opportunities to be positive and encouraging is good for the relationship, especially when it is feeling strained.

6. Control your emotions. It's easy for your temper to flare when your teen is being rude, but don't respond in kind. Remember that you're the adult and he is less able to control his emotions or think logically when he's upset. Count to ten or take some deep breaths before responding. If you are both too upset to talk, hit pause until you've had a chance to calm down.

7. Do things together. Talking isn't the only way to communicate, and during these years it's great if you can spend time doing things you both enjoy, whether it's cooking or hiking or going to the movies, without talking about anything personal. It's important for kids to know that they can be in proximity to you, and share positive experiences, without having to worry that you will pop intrusive questions or call them on the carpet for something.

8. Share regular meals. Sitting down to eat a meal together as a family is another great way to stay close. Dinner conversations give every member of the family a chance to check in and talk casually about sports or television or politics. Kids who feel comfortable talking to parents about everyday things are likely to be more open when harder things come up, too. One rule: no phones allowed.

9. Be observant. It's normal for kids to go through some changes as they mature, but pay attention if you notice changes to her mood, behavior, energy level, or appetite. Likewise, take note if he stops wanting to do things that used to make him happy, or if you notice him isolating himself. If you see a change in your teen's daily ability to function, ask her about it and be supportive (without being judgmental). She may need your help and it could be a sign she needs to talk to a mental health professional.

MISTAKES PARENTS MAKE WITH TEENS

Parenting must change if you wish to keep your relationships strong.

There have been times when parents have been frustrated and we admit it, a bit hurt over our inability to connect with teens. We are guessing our frustrations are shared by other parents who are trying so hard but going about this relationship in all the wrong ways.

Do you speak to your teens as if they are still little kids? Parenting must change if you wish to keep your relationships strong. This includes not only the content but also the tone of conversation. "You need to treat them more like adults than children. Truly listen and heed their point of view, even if you disagree vehemently," says John Duffy, clinical psychologist and author of the "The Available Parent: Radical Optimism for Raising Teens and Tweens." "We all want our point of view respected, and your teen is no different."

Tough as it is, welcoming disagreement in a spirit of humility is foundational. "Mutual respect is so important to teens," says adolescent psychiatrist Meg van Achtenberg. "If you are frustrated that your teen is rolling her eyes, using bad language and talking back, ask yourself if you are treating her with respect. Are you talking to her with the tone you would use to talk to one of your friends?" It's a critical posture change for parents to make, one that can be disorienting.
However, it's crucial, Duffy says, because teens are keen and highly aware people.

Are you treating conversation with them as if it's a chore or obligation? If you are, your teens know it, and it hurts. Teens sniff out adults who pander to them and suffer through dutiful conversation before turning to other adults in a room.

Conversations also should not center on lecturing. "The occasional conversation may be a chore, a bit of a lecture, or a focus on behavior we as parents do not favor. But the lion's share of the discussion has got to be connecting, talking, laughing and sharing," Duffy says.

Maria Coyle, associate head of school for George Washington University Online High School, reminds parents that even after they've built a solid foundation for their kids and prepared them to handle the outside influences of peers and social media, they still have an important role to play in their teens' lives. "Being present for your child, talking with them, noticing things and encouraging them continues to positively affect their growth and development," she says.

The hard work of parenting has changed, but it is certainly not over. Intentional, proactive engagement in our teens' lives is more important than ever, Duffy and Coyle say.

Do you multitask while listening to them? When you're multitasking while your teens are talking to you, it's communicating that they don't warrant your full attention. Van Achtenberg, founder of Capitol Hill Child Psychiatry, urges parents to drop everything if their teens want to talk. "Put down your cellphone, computer, laundry or whatever pressing matters you have, because nothing is more important than hearing out your teenager when he wants to talk."

She points out that evenings and car rides are times when teens are most eager to communicate.

Do you interrupt them? Do you finish their sentences, laugh before they are finished or react in any way before they are done talking? In parents' desperation to relate to their teens, to be cool or to demonstrate energetic engagement, remarks and reactions may easily come out forced and unnatural. Relax. View your teens as good friends, van Achtenberg says. She acknowledges that although a relationship between adult friends won't have the boundaries and consequences present in a parent-teen relationship, "showing respect and kindness toward [your teen] is as essential as it would be toward a friend."

Duffy suggests parents remain silent as much as possible. "My strong bias is to listen more, speak and interrupt less," he says. "Getting to know their world will diminish your parental anxiety."

Do you press them into activities of your choosing? Or do you give them permission to pursue their passions?
Coyle, a licensed school counselor and adolescent behavioral researcher, gets to the heart of it: "This is the time when adolescents question: 'What do I want to do? What do I want to be?' Offering space and support for this exploration allows for a healthy identity to grow. When that space is not offered, an adolescent's identity may not have the room to fully develop."
Instead of mirroring your own hopes and dreams, let your teens take responsibility for their own pursuits. And for teens who lack motivation? Duffy suggests pulling the parent card and insisting that they be involved in something. "Kids need this to balance out the social and academic stressors in their lives and may find their passion through the trial and error of being involved."

Do you try to force the conversation too often? Sometimes, parents try too hard. They want to get kids talking but don't really know how. One idea: Try sharing something from your own day, van Achtenberg says. "The weirdest thing happened at work today, and I couldn't figure out what to do about it' can lead to a conversation in which your teen may be empowered to share advice with you, a wonderful state of affairs for their confidence and your connection. A little vulnerability on your part ('Mom just admitted that she didn't know what to do?') can go a long way."

Look for moments, such as when your younger kids are in bed, to invite your teens to join you in more complex conversations or TV shows that can lead to deep discussions, van Achtenberg suggests.

Coyle says that dinnertime is always a great time to talk, too. There are fewer distractions allowed, permitting conversation to move through the highs and lows of the day. Gathering teens around the dinner table provides an easygoing, nonthreatening setting to get them talking, Coyle says.

Try questions that are open-ended and nonjudgmental. For instance, Coyle suggests: "What was the best thing that happened today? What happened today that you did not like?" Forget the "why" questions or those that can be answered with a yes or no.

Get into their world. Listen to music with them, watch a show they like, or ask them about their favorite social media apps or video games, Duffy suggests. "These conversations provide the goodwill and leverage you will need when things are not going so well. And this is also the good stuff of parenting: truly getting to know and appreciate your kids."

Do they leave the house with you calling out behind them: "Remember, drive slowly! Be safe! Text me!" I tend to worry about my teens, and I know I'm not alone. But when messages of safety are the last thing teens hear every time they leave home, it begins to sound as if you don't trust them.

Offer basic human respect to your teens in these moments, van Achtenberg says, though this doesn't mean allowing your teens to drive if they've demonstrated irresponsibility or to hang out with friends you do not trust.

The point is, timing is everything. Lobbing safety phrases at teens every time they leave the house only dilutes the well-intentioned meaning behind them. The time to discuss road safety, for example, should not be when teens are flying out the door. "Teachable moments will arise. Utilize those opportunities to have more in-depth conversations," Coyle suggests.

So, bite your tongue. Instead, tell your teens you love them, and, after they depart, murmur your prayers and send positive vibes in their direction.

In contrast to when your teens were children, the way you relate to them may be fundamentally uncomfortable. Providing your teens with greater autonomy as they grow essentially means learning a new way to care for them. The learning curve will be well worth your time and effort, however, rewarding you with natural camaraderie.

HEALTHY RELATIONSHIPS FOR TEENS

The sex talk is not enough: How can parents teach teens about healthy relationships, especially boys?

Parents fret for years about having "the talk" with their kids. That talk, of course, being about sex. But one thing that is getting very lost in those conversations is how to have a healthy romantic relationship. It's not enough to have the sex talk, we have to have the love talk, too. Without it, we risk our kids being in abusive, manipulative relationships, or missing out on a truly wonderful aspect of life.

According to a report released today by Harvard Graduate School of Education's Making Caring Common Project, parents worry a great deal about the hookup culture, but ignore the fact that young people are unprepared to learn how to love and develop caring, healthy romantic relationships.

"This whole area has been terribly neglected," says Richard Weissbourd, a Harvard psychologist who runs the Making Caring Common project. Without conversations about healthy relationships, parents are also neglecting to teach their children about misogyny and sexual harassment. "Adults seem not to be facing it squarely. It's concerning," Weissbourd adds.

If parents think kids don't want to hear it from them, they should reconsider: 70 percent of the 18- to 25-year-olds who responded to the report's survey said they wanted more information from parents about some emotional aspect of a romantic relationship. And 65 percent said they wanted guidance about it in a sex or health class at school. But both parents and educators seem to focus on abstinence, how not to get pregnant or how to avoid a sexually transmitted disease. In doing so, parents are missing out on having important conversations about how to love and be loved.

Here, Weissbourd and the report's authors offer five ways to teach kids and young adults about healthy relationships:

Be a romantic philosopher

Why? Young people and adults mean very different things when they say they're in love. Because our understandings of love are vague and varied, young people may confuse love with infatuation, lust, idolization or obsession. They may think, for example, that they are in love with someone because they can't stop thinking about them. Or they may confuse love with the boost in self-esteem they experience when someone is romantically interested in them.

Try this: Speak with your teen about the many forms of love. Explain what you mean when you say that you are in love with someone. Let your child understand that they may define being in love differently than someone else and that there is no right definition of being in love. But there are ways of knowing whether intense feelings for someone else are likely to lead to healthy or unhealthy romantic relationships. Explore with your teen why and how love can be deeply meaningful and change the course of our lives.

Also: Ask your child how they think about different types of intense feelings toward someone. Talk about how people can be attracted to, or preoccupied with, other people for a range of positive and negative reasons, and discuss the importance of understanding why your teen might be attracted to someone else. Are they attracted to someone at least partly because they're kind, generous and honest? Or are they attracted to someone because that person is elusive, seems unattainable or mistreats them in some way? Discussing these questions can give them tools for determining whether a relationship is likely to be healthy or unhealthy.

Talk about the markers of healthy and unhealthy relationships

Why? Teens may not know whether they're in a healthy or unhealthy relationship. They also may be unsure if their worries, feelings of disappointment or criticisms of their partner are normal.

Try this: Examples of both healthy and unhealthy relationships are everywhere. Talk to your teen about couples you both know, and representations of relationships in the media. Which are healthy? Which are harmful? Why? If your teen is in a relationship, you might ask whether it makes them more or less self-respecting, hopeful, caring and generous.

Talk about the skills needed to maintain healthy relationships

Why? Maintaining healthy relationships requires a range of skills, including the ability to communicate honestly and effectively, to jointly solve problems, to measure anger and to be generous. Healthy relationships also benefit from being able to take someone else's perspective in a deep way and to step back and view the relationship and its dynamics, strengths and challenges.

Try this: Discuss with your child various examples of caring, vibrant relationships. These examples might be relatives or friends who you think have mature romantic relationships, or could be couples portrayed in books, television, movies. You might watch with your teen the compelling marriages depicted in shows such as "This Is Us" and "Friday Night Lights."

Consider sharing lessons from your own relationships

Why? We can mine our experiences for insights about mature and immature love, and why relationships do and don't work. Teens are often interested in our experiences, partly because they're sorting out how they're like or unlike us.

Try this: Think about what your relationships have taught you. What was healthy about them? Unhealthy? What attitudes or behaviors would you change if you could? Share with your teen any lessons you have learned about the skills, attitudes and sensitivities it takes to maintain a healthy relationship.

Engage young people in ethical questions connected to romantic and sexual relationships

Why? High school and college students enthusiastically plunge into ethical questions about romantic relationships: *What do I do if I know my friend is cheating on his girlfriend who is also my friend? Is it exploitation when a senior hook up with a first year?* Reflecting on such questions can help young people develop better relationships, but also help them develop complex thinking and problem-solving skills, and learn to ethically reason when dealing with conflicting loyalties, and take up questions about human rights and dignity.

Try this: Together with your child, puzzle through answers to ethical questions. Start by listening to how your teen would answer these questions, then share your own thoughts. Often there is not one right answer. Consider how to resolve these dilemmas in ways that are as fair, honest and caring for all people involved.

PUBERTY BOYS – ALCOHOL AND DRUGS

Strategies to help teenagers make good decisions

Helping teenagers make good decisions about drinking and drugs can be a huge challenge for parents—especially for those who are uncomfortable about setting limits that they distinctly remember violating in their own teenage years. But in an environment where alcohol and pot are ubiquitous, research shows that clear parental direction helps kids reel in substance use and dangerous behavior.

What do you hear most from parents, in terms of what they are worried about?
The first thing that comes to my mind is the difficulties that parents have with limit-setting, and that they are often looking for reassurance or permission, for someone else to tell them that yes, indeed, setting limits is something that they can do and should do—whether it's setting a curfew time, or saying no to a party that is not going to be supervised, or laying out expectations that there be no drinking until they're 21.

You think it is a mistake for parents to acknowledge that they expect kids will experiment?
Talking about "experimentation" tends to be a very slippery slope. Parents have one idea about what it means, and kids have another idea about what it means. And some kids who get the idea that experimentation is okay will rationalize any kind of drinking that they're doing as experimentation.

The word means something different to parents and kids?
When parents use it, I think they often rationalize: "They're just going to try alcohol, or maybe they're going to get drunk once, or maybe they will even try marijuana." But teens can use it as a catch-all phrase for partying—regularly drinking to get drunk, regularly using marijuana. Research shows that if parents express expectations for experimentation, their kids will use, and the continuum of use tends to be more dramatic than kids whose parents lay out expectations for no experimentation.

The law aside, why is drinking so problematic for teens?
Parents should understand that their children's brains are not fully formed until they are in their mid-twenties. So if adolescents use alcohol, they are much more likely to feel the adverse effects, and they get drunker on less alcohol, and much quicker. Parents forget that their kids' bodies are not like theirs, even though they may look like similar, because they're growing up fast. Parents may be able to exercise moderation while their children may not be able to yet. Parents and kids are starting to understand there is some biological basis for why drinking is postponed until 21.

Are parents worried that they will stunt their kids' social lives if they are strict about drinking?
We hear from parents, "I don't want my child to go to college and suddenly be faced with a keg party and not know what to do"—or, you could even say, not be popular, not fit in, not have the social skills that should be developed in upper school. So it's kind of this subtle message of, "I want my child to be popular. I want my child to be able to go to a party. I want my child to experiment, perhaps, like I experimented at that age"—again, with the misperception of what really is going on, when there's binge drinking, or pre-gaming, or five shots, not one beer. It's quite different now.

Is there social pressure on parents to go along with the drinking?
There is actually peer pressure among parents. One of the things that I've observed at some schools is that parents are even reticent to say that they don't want their child to drink, because they feel that it might make them unpopular with other parents. Because then it shifts the expectations for the community. If there are people who are vocal about not wanting alcohol to be served, and they make everyone accountable for that, it can be very tricky territory for people to enter.

Other parents do not appreciate being told about their kid's drinking?
Many parents say, "I would really want anybody to call me, if you thought my child was doing something harmful." But it's exactly the opposite when that actually has transpired. I think people are usually caught off-guard, blindsided, and embarrassed, and don't know what to do with the information. It may have a positive impact down the line, but the bearer of the bad news doesn't necessarily get the positive feedback. There may be a defensive reaction: "Well, lots of kids are drinking. Why are you singling me out?"

If a parent is not comfortable saying no to any experimentation, is there a limit that you think works, short of abstinence?
We would say that the way to talk to teenagers about this is to be very black and white. Keep that leash really short. When it comes to substance abuse, you don't want to give them any room, or any permission, to be out drinking with their underage friends. Even if you suspect that it may happen, you want to give the message that you do not expect it.

You want to voice the expectation that they will not drink, but you also want them to call you if they end up in a bad situation. This is a really tricky line for parents to tread. We find it helps if you express something along the lines of, "You know that we do not want you to drink, and we expect that you will make good decisions. But if anything happens that you are uncomfortable with, please, please, call us. We will help you, and it's much more important that you ask for help than that you may have made a stupid decision."

What's an appropriate consequence for breaking the rules on this?
It's best to let the consequences come out of the experience itself, and be directly related to it. Because the issue is kids being able to handle being out with their peers in a sober way, most often, parents settle on some kind of grounding. But we do encourage parents to wait, and make that decision on a case-by-case basis.

Some parents have an amnesty policy (I do with my children), where if you have the good judgment to call for help, you are not punished for the actual drinking. Every parent has to make their own decision on this, but I've generally found it be successful, because it emphasizes the relationship, rather than the behavior. Kids respond to this, and take away the impression that their parents are reasonable people who care, first and foremost, about their safety.

Do you think that parents are reluctant to set limits because they don't want their kids to lie to them?
That's very often a rationalization, or a fear, that parents have.
Along those lines, one of the other primary points that we share with parents is not to talk about their own past use, as a parent. It's a rule of thumb to really think twice if your child is asking you a direct question about your own use, because it's almost guaranteed that any question they're asking you is really about them, and the choices that they are facing.

What do you tell them to say?
We suggest that parents really let their children know that whatever happened in the past in the parents' lives probably has very little relevance to this particular time and place. "So much has changed, and what we're really interested in is making sure that you're safe and sound. And one of the ways I want to make sure that's the case is by expecting you not to drink." It's a chance to reiterate the rules. And certainly to give them lots of room to talk about it, about what they're being exposed to.

Parents are nervous about these conversations. What about kids?
I have been meeting a lot of upper school students in the past month or so, and asking them, "Would you all want to know where your parents stand on alcohol and drugs, or would you not want to know it?" Every kid, even the kids who have admitted using, says, "I'd rather know where my parents stand." And they do want to know. I ask them, "Do your parents influence the choices you make?" And I remember this one student said, "Yeah, I know

that my dad would get really mad at me if I came home drunk. So I really have to think twice, if I'm at a party: Do I pretend, or do I just not drink?" Some kids say, "I know my parent doesn't want me to drink, so I'll go somewhere else." That's the minority. But every single kid, in the five schools I went to, said, "I'd rather know." I mean, not one kid said "I don't want to know, because if I know, I'll feel bad." And it was great to hear that."

It is very hard—and I've been through this myself, as a parent—it is very hard to sort of own your parental authority. And you also have this deep desire to be straight with your kids. And it's so easy to get mixed up. But increasingly, I believe that kids want parents to be parents, even the most rebellious child.

Probably particularly the most rebellious child. They really want you to be a parent. And, you know, that does not necessarily always mean being completely straight with them.

Each of us has an experience of a teenager telling us, "Well, when my mother told me that she smoked pot, I was horrified! I didn't really want that image in my head, of my mom getting high."

That is an interesting phrase, "I do not want to have that image in my head."
They idealize parents so much. And you want the authority in your life to be a certain way.

ABOUT THE AUTHOR

Jameka Watkins is the New York Times bestselling author of The Joy of Leaving Your Sh*t All Over the Place. She is also the author of Cocktails for Drinkers, Poetry from Scratch and the novel Afloat. Her work has appeared in the Atlantic, Teen Vogue and on BBC Radio 4.